Liaison Functions; Chemical Engineering & the Law Forum; Leadership Development - Professional Skills to Catalyze your Career 2023

Topical Conference at the 2023 AIChE Spring Meeting and 19th Global Congress on Process Safety

Houston, Texas, USA
12-16 March 2023

ISBN: 978-1-7138-7931-2

Printed from e-media with permission by:

Curran Associates, Inc.
57 Morehouse Lane
Red Hook, NY 12571

Some format issues inherent in the e-media version may also appear in this print version.

Printed with permission by Curran Associates, Inc. (2023)

For permission requests, please contact AIChE
at the address below.

AIChE
120 Wall Street, FL 23
New York, NY 10005-4020

Phone: (800) 242-4363
Fax: (203) 775-5177

www.aiche.org

Additional copies of this publication are available from:

Curran Associates, Inc.
57 Morehouse Lane
Red Hook, NY 12571 USA
Phone: 845-758-0400
Fax: 845-758-2633
Email: curran@proceedings.com
Web: www.proceedings.com

TABLE OF CONTENTS

29a Protecting Your Intellectual Property in Joint Ventures

Paul Townsend, Lauren Dowty

As industries grow and technology develops, companies look to every potential avenue to gain and maintain a competitive advantage in the marketplace. One way of doing this is to enter into joint ventures with other companies. Joint ventures allow companies to combine forces to bring a product or service to the market, often in ways neither company has done before. Critical to the success of a joint venture is the sharing of ideas between the different companies. And while in many instances joint ventures go smoothly with no issues between the parties, everyone should be prepared to discuss the handling and ownership of intellectual property (IP) shared or developed during the pendency of the joint venture.

This presentation will provide an overview of the basics of IP, the IP risks of joint ventures, the protection and maintenance of IP during a joint venture, important considerations relating to ownership of IP developed during a joint venture, and other considerations for companies entering into joint ventures. Speakers will also highlight case studies addressing joint ventures and whether the companies took sufficient precautions.

29b IP Considerations for Industrial Processes

Michael DePierro, John Gohndrone

There are many important considerations for protecting intellectual property for industrial process technology. Understanding the patent landscape for a given technology is critical for guiding an effective strategy for patenting. Thorough prior art searching is necessary for defining the novel aspects of a technology and understanding the scope available for patenting. This assessment can avoid writing patent applications disclosing technology more broadly than can be protected. It helps define claims sufficiently narrow to have a high probability of granting. This approach also provides indications of how likely it is that others will file patents for similar technology. This presentation will cover how these considerations lead to understanding of the potential value of a patent relative to other options for protecting IP for process technology.

109c Mindful Methods for Success

Lisa Liszcz

Mindful Methods for Success
Stop Undermining Yourself and Become Your Own Best Cheerleader

The most effective way to have success in your career is to learn one thing that does not require you to take a course or read a book.
By eliminating something you may do regularly, you can significantly upgrade your career and life while saving time and energy.
One key to creating the career and life that you desire is to spend 10 minutes a day on a simple strategy that few people about. It's easy, painless, and you can do it anywhere.
Lisa M. Liszcz, Ph.D. works with companies on organizational growth and achieving ambitious goals. Dr. Liszcz has found that inclusiveness and valuing diversity are critical for the growth and sustainability of organizations. When Dr. Liszcz coaches professionals, she helps them find new, mindful methods to achieve their goals and be successful.
In her powerful presentation, 'Mindful Methods for Success' Lisa shares three secrets with her audiences:
People who sign up to attend the 'Mindful Methods for Success' presentation are invited to take Dr. Liszcz's 'Self-Esteem' Quiz beforehand. Dr. Liszcz will help participants understand which mindful methods they can use to boost self-esteem and experience more success at work and in life.
This presentation is based on Dr. Liszcz's Amazon.com best-seller, FE Feminine Emergence.
Schedule Lisa to present to your group today and energize your life!

23a Rock the podium: Masterclass on delivering virtual and in-person presentations

Alaina Levine

In this conversation, we will discuss the essentials to delivering an excellent, enthusiastic presentation, whether it is live or via virtual platforms. The topic is applicable to company speeches, job talks, colloquia, conference oral and poster presentations, and any other type of presentations that you may give. We will discuss key tactics that will enable your success as you speak, including how to convey passion, enthusiasm, knowledge, and expertise, how to leverage the parameters of digital platforms like zoom, and how to utilize storytelling to engage the audience. You will emerge with clear direction as to how you can be a stronger, more confident speaker.

39a Create Your Unicorn Career: Innovatively design your customized career in any ecosystem

Alaina Levine

A Unicorn Career is a customized career that you create for yourself, one that is aligned with your authenticity 100% of the time and brings you joy, meaning, and money. Anyone can create a Unicorn Career (and they can be launched in organizations and in professions that already exists!), but it takes bravery, data collection, process development, and effective marketing and networking. It also requires one to leverage skills in entrepreneurship and intrepreneurship, to identify a pain point that you are uniquely qualified to alleviate. In this session, we will discuss the processes by which you can plan and orchestrate a Unicorn Career while we focus on two types of data collection: internal information about your abilities, interests, and objectives, obtained through assessments, and external information about opportunities, career paths, and ecosystems, which we obtain through networking. We will also address the importance of self-promotion and why and how marketing is a key component of advancing your career. You will emerge with a how to list of tactics and strategies you can take literally NOW to launch the career of your dreams (even in challenging times). Your Unicorn Career is not only possible, but probable and realistic. Let's move you forward!

154a Reynolds number to data frames: retraining your mind to think about analytics – Panel Discussion

Joshua Holtz

The journey to data analytics from chemical engineering is one that can feel unachievable, but with a little bit of help the jump to becoming a data scientist can be within your reach. Although no multiple paths will look the same come listen in on some of the ways our panelists have learned to harness their chemical engineering background to transition to data science and in the process make unique contributions to their organizations.

77a Utilizing Science, Collaboration, and Urgency to Set and Achieve ESG goals

Gayle Schuller

- Exploring approaches to engage a broad stakeholder community in developing sustainability strategies, frameworks, and goals consistent with a unique corporate identity and culture.
- Highlighting 3M's Strategic Sustainability Framework including three pillars: Science for Circular, Science for Community and Science for Community.
- Demonstrating how to embed science into the heart of sustainability goals and corporate culture.
- Discussing how to raise ambitions and update goals based on new information.
- Showcasing how to apply a 'math with a path' approach to build confidence and accelerate the pace of carbon emissions, plastic and water reductions through new innovations and technology; and collaborative efforts to help achieve goals even faster.

AUTHOR INDEX

NOTES

NOTES